Cram101 Textbook Outlines to accompany:

Experiments in Physical Chemistry

Carl W. Garland, David P. Shoemaker, Joseph W. Nibler, 8th Edition

A Cram101 Inc. publication (c) 2011.

PRACTICE EXAMS.

Get all of the self-teaching practice exams for each chapter of this textbook at
www.Cram101.com and ace the tests. Here is an example:

Chapter 1

Experiments in Physical Chemistry
Carl W. Garland, David P. Shoemaker, Joseph W. Nibler, 8th Edition,
All Material Written and Prepared by Cram101

I WANT A BETTER GRADE.   Items 1 - 50 of 100. ▶

1 _____, in chemistry, is a physical or chemical phenomenon or a process in which atoms, molecules liquid or solid

material. This is a different process from adsorption, since the molecules are taken up by the volume, not by surface.

A more general term is sorption which covers adsorption, _____, and ion exchange.

◎ Absorption   ◎ Ab initio multiple spawning

◎ ABCN   ◎ ABINIT

2 In physics, _____ is the process by which the energy of a photon is released by another entity, for example, by

an atom whose electrons make a transition between two electronic energy levels. The emitted energy is in the form of

a photon.

The emittance of an object quantifies how much light is emitted by it.

◎ Emission   ◎ E1cB elimination reaction

◎ Earle K. Plyler Prize for Molecular   ◎ Earth battery

Spectroscopy

3 _____ is a luminescence that is mostly found as an optical phenomenon in cold bodies, in which the molecular

absorption of a photon triggers the emission of a photon with a longer (less energetic) wavelength, though a shorter

wavelength emission is sometimes observed from multiple photon absorption. The energy difference between the

absorbed and emitted photons ends up as molecular rotations, vibrations or heat. Sometimes the absorbed photon is

in the ultraviolet range, and the emitted light is in the visible range, but this depends on the absorbance curve and

Stokes shift of the particular fluorophore

You get a 50% discount for the online exams. Go to **Cram101.com**, click Sign Up at
the top of the screen, and enter DK73DW6081 in the promo code box on the
registration screen. Access to Cram101.com is $4.95 per month, cancel at any time.

With Cram101.com online, you also have access to extensive reference material.

You will nail those essays and papers. Here is an example from a Cram101 Biology text:

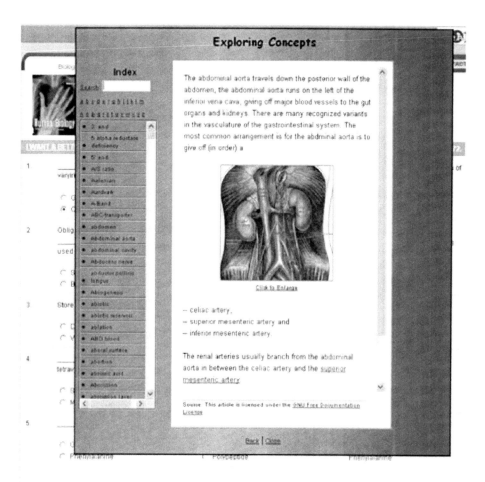

Visit **www.Cram101.com**, click Sign Up at the top of the screen, and enter DK73DW6081 in the promo code box on the registration screen. Access to www.Cram101.com is normally $9.95 per month, but because you have purchased this book, your access fee is only $4.95 per month, cancel at any time. Sign up and stop highlighting textbooks forever.

Learning System

Cram101 Textbook Outlines is a learning system. The notes in this book are the highlights of your textbook, you will never have to highlight a book again.

How to use this book. Take this book to class, it is your notebook for the lecture. The notes and highlights on the left hand side of the pages follow the outline and order of the textbook. All you have to do is follow along while your instructor presents the lecture. Circle the items emphasized in class and add other important information on the right side. With Cram101 Textbook Outlines you'll spend less time writing and more time listening. Learning becomes more efficient.

Cram101.com Online

Increase your studying efficiency by using Cram101.com's practice tests and online reference material. It is the perfect complement to Cram101 Textbook Outlines. Use self-teaching matching tests or simulate in-class testing with comprehensive multiple choice tests, or simply use Cram's true and false tests for quick review. Cram101.com even allows you to enter your in-class notes for an integrated studying format combining the textbook notes with your class notes.

Visit **www.Cram101.com**, click Sign Up at the top of the screen, and enter **DK73DW6081** in the promo code box on the registration screen. Access to www.Cram101.com is normally $9.95 per month, but because you have purchased this book, your access fee is only $4.95 per month. Sign up and stop highlighting textbooks forever.

Experiments in Physical Chemistry
Carl W. Garland, David P. Shoemaker, Joseph W. Nibler, 8th

CONTENTS

Absorption	Absorption, in chemistry, is a physical or chemical phenomenon or a process in which atoms, molecules liquid or solid material. This is a different process from adsorption, since the molecules are taken up by the volume, not by surface. A more general term is sorption which covers adsorption, Absorption, and ion exchange.
Emission	In physics, Emission is the process by which the energy of a photon is released by another entity, for example, by an atom whose electrons make a transition between two electronic energy levels. The emitted energy is in the form of a photon. The emittance of an object quantifies how much light is emitted by it.
Fluorescence	Fluorescence is a luminescence that is mostly found as an optical phenomenon in cold bodies, in which the molecular absorption of a photon triggers the emission of a photon with a longer (less energetic) wavelength, though a shorter wavelength emission is sometimes observed from multiple photon absorption. The energy difference between the absorbed and emitted photons ends up as molecular rotations, vibrations or heat. Sometimes the absorbed photon is in the ultraviolet range, and the emitted light is in the visible range, but this depends on the absorbance curve and Stokes shift of the particular fluorophore.
Gas	In physics, a Gas is a state of matter, consisting of a collection of particles (molecules, atoms, ions, electrons, etc.) without a definite shape or volume that are in more or less random motion. Due to the electronic nature of the aforementioned particles, a "force field" is present throughout the space around them.
Quenching	Quenching refers to any process which decreases the fluorescence intensity of a given substance. A variety of processes can result in Quenching, such as excited state reactions, energy transfer, complex-formation and collisional Quenching. As a consequence, Quenching is often heavily dependent on pressure and temperature.
Resonance fluorescence	Resonance fluorescence is fluorescence from an atom or molecule in which the light emitted is at the same frequency as the light absorbed. In Resonance fluorescence, a photon is absorbed, causing an electron to jump to a higher level from which, after a delay, it falls back to its original level, emitting a photon having the same energy as the one absorbed. The emission direction is random.
Density	The Density of a material is defined as its mass per unit volume. The symbol of Density is ρ ">rho.) Mathematically: $$\rho = \frac{m}{V}$$

where:

ρ is the Density,
m is the mass,
V is the volume.

Density of air

The Density of air, ρ , is the mass per unit volume of Earth"s atmosphere, and is a useful value in aeronautics and other sciences. Air density decreases with increasing altitude, as does air pressure. It also changes with variances in temperature or humidity.

Morse potential

The Morse potential is a convenient model for the potential energy of a diatomic molecule. It is a better approximation for the vibrational structure of the molecule than the quantum harmonic oscillator because it explicitly includes the effects of bond breaking, such as the existence of unbound states. It also accounts for the anharmonicity of real bonds and the non-zero transition probability for overtone and combination bands.

Molecular orbital

In chemistry, a Molecular orbital is a mathematical function that describes the wave-like behavior of an electron in a molecule. This function can be used to calculate chemical and physical properties such as the probability of finding an electron in any specific region. The use of the term "orbital" was first used in English by Robert S. Mulliken in 1925 as the English translation of Schrödinger"s use of the German word, "Eigenfunktion".

Molecular orbital theory

In chemistry, Molecular orbital theory is a method for determining molecular structure in which electrons are not assigned to individual bonds between atoms, but are treated as moving under the influence of the nuclei in the whole molecule. In this theory, each molecule has a set of molecular orbitals, in which it is assumed that the molecular orbital wave function ψ_f may be written as a simple weighted sum of the n constituent atomic orbitals χ_i, according to the following equation:

$$\psi_j = \sum_{i=1}^{n} c_{ij}\chi_i$$

The c_{ij} coefficients may be determined numerically by substitution of this equation into the Schrödinger equation and application of the variational principle. This method is called the linear combination of atomic orbitals approximation and is used in computational chemistry.

Triple point

In thermodynamics, the Triple point of a substance is the temperature and pressure at which three phases (for example, gas, liquid, and solid) of that substance coexist in thermodynamic equilibrium. For example, the Triple point of mercury occurs at a temperature of –38.8344 °C and a pressure of 0.2 mPa.

In addition to the Triple point between solid, liquid, and gas, there can be Triple point s involving more than one solid phase, for substances with multiple polymorphs.

Compressibility	In thermodynamics and fluid mechanics, Compressibility is a measure of the relative volume change of a fluid or solid as a response to a pressure (or mean stress) change. $$\beta = -\frac{1}{V}\frac{\partial V}{\partial p}$$ where V is volume and p is pressure. The above statement is incomplete, because for any object or system the magnitude of the Compressibility depends strongly on whether the process is adiabatic or isothermal.
Compressibility factor	The Compressibility factor is a useful thermodynamic property for modifying the ideal gas law to account for the real gas behaviour. In general, deviations from ideal behavior become more significant the closer a gas is to a phase change, the lower the temperature or the larger the pressure. Compressibility factor values are usually obtained by calculation from equations of state (EOS), such as the virial equation which take compound specific empirical constants as input.
Equation of state	In physics and thermodynamics, an Equation of state is a relation between state variables. More specifically, an Equation of state is a thermodynamic equation describing the state of matter under a given set of physical conditions. It is a constitutive equation which provides a mathematical relationship between two or more state functions associated with the matter, such as its temperature, pressure, volume, or internal energy.
Helium	Helium is the chemical element with atomic number 2, and is represented by the symbol He. It is a colorless, odorless, tasteless, non-toxic, inert monatomic gas that heads the noble gas group in the periodic table. Its boiling and melting points are the lowest among the elements and it exists only as a gas except in extreme conditions.
Boiling	Boiling, a type of phase transition, is the rapid vaporization of a liquid, which typically occurs when a liquid is heated to its Boiling point, the temperature at which the vapor pressure of the liquid is equal to the pressure exerted on the liquid by the surrounding environmental pressure. Thus, a liquid may also boil when the pressure of the surrounding atmosphere is sufficiently reduced, such as the use of a vacuum pump or at high altitudes. Boiling occurs in three characteristic stages, which are nucleate, transition and film Boiling.
Boiling point	The Boiling point of an element or a substance is the temperature at which the vapor pressure of the liquid equals the environmental pressure surrounding the liquid. A liquid in a vacuum environment has a lower Boiling point than when the liquid is at atmospheric pressure. A liquid in a high pressure environment has a higher Boiling point than when the liquid is at atmospheric pressure.
Pressure	Pressure is the force per unit area applied in a direction perpendicular to the surface of an object. Gauge Pressure is the Pressure relative to the local atmospheric or ambient Pressure. Pressure is an effect which occurs when a force is applied on a surface.

Adiabatic process	In thermodynamics, an Adiabatic process or an isocaloric process is a thermodynamic process in which no heat is transferred to or from the working fluid. The term "adiabatic" literally means impassable, coming from the Greek roots á¼€- , διá½°- ("through"), and βαá¿–νειν ("to pass"); this etymology corresponds here to an absence of heat transfer. Conversely, a process that involves heat transfer (addition or loss of heat to the surroundings) is generally called diabatic.
Van der Waals equation	The Van der Waals equation is an equation of state for a fluid composed of particles that have a non-zero size and a pairwise attractive inter-particle force (such as the van der Waals force.) It was derived by Johannes Diderik van der Waals in 1873, based on a modification of the ideal gas law, who received the Nobel prize in 1910 for "his work on the equation of state for gases and liquids". The equation approximates the behavior of real fluids, taking into account the nonzero size of molecules and the attraction between them.
Heat	In physics and thermodynamics, Heat is the process of energy transfer from one body or system to another due to a difference in temperature. In thermodynamics, the quantity TdS is used as a representative measure of the (inexact) Heat differential δQ, which is the absolute temperature of an object multiplied by the differential quantity of a system"s entropy measured at the boundary of the object.

A related term is thermal energy, loosely defined as the energy of a body that increases with its temperature. |
| Inversion | In meteorology, an Inversion is a deviation from the normal change of an atmospheric property with altitude. It almost always refers to a temperature Inversion, i.e., an increase in temperature with height, or to the layer (Inversion layer) within which such an increase occurs.

An Inversion can lead to pollution such as smog being trapped close to the ground, with possible adverse effects on health. |
| Inversion temperature | The Inversion temperature in thermodynamics and cryogenics is the critical temperature below which a non-ideal gas (all gases in reality) that is expanded at constant enthalpy will experience a temperature decrease, and above which will experience a temperature increase. This temperature change is known as the Joule-Thomson effect, and is exploited in the liquefaction of gases.

The Joule-Thomson effect cannot be described in the theory of ideal gases, in which interactions between particles are ignored. |
| Adsorption | Adsorption is the accumulation of atoms or molecules on the surface of a material. This process creates a film of the adsorbate (the molecules or atoms being accumulated) on the adsorbent"s surface. It is different from absorption, in which a substance diffuses into a liquid or solid to form a solution. |

Equipartition theorem	In classical statistical mechanics, the Equipartition theorem is a general formula that relates the temperature of a system with its average energies. The Equipartition theorem is also known as the law of equipartition, equipartition of energy, or simply equipartition. The original idea of equipartition was that, in thermal equilibrium, energy is shared equally among all of its various forms; for example, the average kinetic energy in the translational motion of a molecule should equal the average kinetic energy in its rotational motion.
Kinetic theory	Kinetic theory attempts to explain macroscopic properties of gases, such as pressure, temperature by considering their molecular composition and motion. Essentially, the theory posits that pressure is due not to static repulsion between molecules, as was Isaac Newton"s conjecture, but due to collisions between molecules moving at different velocities. Kinetic theory is also known as the Kinetic-Molecular Theory or the Collision Theory or the Kinetic-Molecular Theory of Gases.
Transport phenomena	In physics, chemistry, biology and engineering, a transport phenomenon is any of various mechanisms by which particles or quantities move from one place to another. The laws which govern transport connect a flux with a "motive force". Three common examples of Transport phenomena are diffusion, convection, and radiation.
Collision frequency	Collision frequency is defined in chemical kinetics, in the background of theoretical kinetics, as the average number of collisions between reacting molecules per unit of time. Its symbol is Z. Unlike the preexponential factor, A, which is an empirical magnitude, Z is a theoretical magnitude, calculated ab initio. The closer A and Z are, the better a given theory of chemical kinetics is.
Effusion	In chemistry, Effusion is the process in which individual molecules flow through a hole without collisions between molecules. This occurs if the diameter of the hole is considerably smaller than the mean free path of the molecules. According to Graham"s law, the rate at which gases effuse (i.e., how many molecules pass through the hole per second) is dependent on their molecular weight; gases with a lower molecular weight effuse more quickly than gases with a higher molecular weight.
Thermal conductivity	In physics, Thermal conductivity, k, is the property of a material that indicates its ability to conduct heat. It appears primarily in Fourier"s Law for heat conduction. First, we define heat conduction, H: $$H = \frac{\Delta Q}{\Delta t} = kA\frac{\Delta T}{x}$$ where $\frac{\Delta Q}{\Delta t}$ is the rate of heat flow, k is the Thermal conductivity, A is the total cross sectional area of conducting surface, ΔT is temperature difference, and x is the thickness of conducting surface separating the 2 temperatures.

Viscometer	A Viscometer is an instrument used to measure the viscosity of a fluid. For liquids with viscosities which vary with flow conditions, an instrument called a rheometer is used. Viscometer s only measure under one flow condition.
Activation	Activation in (bio-)chemical sciences generally refers to the process whereby something is prepared or excited for a subsequent reaction. In chemistry, Activation of molecules is required for a chemical reaction to occur. The phrase energy of Activation refers to the energy the reactants must acquire before they can successfully react with each other to produce the products, that is, to reach the transition state.
Activation energy	In chemistry, Activation energy is a term introduced in 1889 by the Swedish scientist Svante Arrhenius, that is defined as the energy that must be overcome in order for a chemical reaction to occur. Arrhenius" research was a follow up of the theories of reaction rate by Serbian physicist Nebojsa Lekovic. Activation energy may also be defined as the minimum energy required to start a chemical reaction.
Calorimetry	Calorimetry is the science of measuring the heat of chemical reactions or physical changes. Calorimetry involves the use of a calorimeter. The word Calorimetry is derived from the Latin word calor, meaning heat.
Enthalpy	In thermodynamics and molecular chemistry, the Enthalpy is a thermodynamic property of a fluid. It can be used to calculate the heat transfer during a quasistatic process taking place in a closed thermodynamic system under constant pressure. Enthalpy H is an arbitrary concept but the Enthalpy change ΔH is more useful because it is equal to the change in the internal energy of the system, plus the work that the system has done on its surroundings.
Thermochemistry	In thermodynamics and physical chemistry, Thermochemistry is the study of the energy evolved or absorbed in chemical reactions and any physical transformations, such as melting and boiling. Thermochemistry, generally, is concerned with the energy exchange accompanying transformations, such as mixing, phase transitions, chemical reactions, and including calculations of such quantities as the heat capacity, heat of combustion, heat of formation, enthalpy, and free energy. Thermochemistry rests on two generalizations: 1. Lavoisier and Laplace"s law (1782): the heat exchange accompanying a transformation is equal and opposite to the heat exchange accompanying the reverse transformation. 2. Hess"s law (1840): the heat exchange accompanying a transformation is the same whether the process occurs in one or both steps Both of these statements preceded the first law of thermodynamics (1850) and helped in its formulation.

Lavoisier, Laplace and Hess also investigated specific heat and latent heat, although it was Joseph Black who made the most important contributions to the development of latent energy changes.

Joule heating

Joule heating is the process by which the passage of an electric current through a conductor releases heat. It was first studied by James Prescott Joule in 1841. Joule immersed a length of wire in a fixed mass of water and measured the temperature rise due to a known current flowing through the wire for a 30 minute period.

Acid

An Acid is traditionally considered any chemical compound that, when dissolved in water, gives a solution with a hydrogen ion activity greater than in pure water, i.e. a pH less than 7.0. That approximates the modern definition of Johannes Nicolaus Brønsted and Martin Lowry, who independently defined an Acid as a compound which donates a hydrogen ion (H^+) to another compound (called a base.) Common examples include acetic Acid and sulfuric Acid (used in car batteries.)

Solution

In chemistry, a Solution is a homogeneous mixture composed of two or more substances. In such a mixture, a solute is dissolved in another substance, known as a solvent. Gases may dissolve in liquids, for example, carbon dioxide or oxygen in water.

Energy levels

A quantum mechanical system or particle that is bound, confined spatially, can only take on certain discrete values of energy, as opposed to classical particles, which can have any energy. These values are called Energy levels. The term is most commonly used for the Energy levels of electrons in atoms or molecules, which are bound by the electric field of the nucleus.

Chemical equilibrium

In a chemical process, Chemical equilibrium is the state in which the chemical activities or concentrations of the reactants and products have no net change over time. Usually, this would be the state that results when the forward chemical process proceeds at the same rate as their reverse reaction. The reaction rates of the forward and reverse reactions are generally not zero but, being equal, there are no net changes in any of the reactant or product concentrations.

Gibbs-Duhem equation

The Gibbs-Duhem equation in thermodynamics describes the relationship between changes in chemical potential for components in a thermodynamical system :

$$\sum_{i=1}^{I} N_i \mathrm{d}\mu_i = -S\mathrm{d}T + V\mathrm{d}p$$

where N_i is the number of moles of component i, $d\mu_i$ the incremental increase in chemical potential for this component, S the entropy, T the absolute temperature, V volume and P the pressure. It shows that in thermodynamics intensive properties are not independent but related, making it a mathematical statement of the state postulate. When pressure and temperature are variable, only $I - 1$ of I components have independent values for chemical potential and Gibbs" phase rule follows.

Mole fraction

In chemistry, Mole fraction x (also, and more correctly, known as the amount fraction) is a way of expressing the composition of a mixture. The Mole fraction of each component i is defined as its amount of substance n_i divided by the total amount of substance in the system, n

$$x_i \overset{\text{def}}{=} \frac{n_i}{n}$$

where

$$n = \sum_i n_i$$

The sum is over all components, including the solvent in the case of a chemical solution. As an example, if a mixture is obtained by dissolving 10 moles of sucrose in 90 moles of water, the Mole fraction of sucrose in that mixture is 0.1.

Chemical potential

Chemical potential, symbolized by μ, is a quantity first described by the American engineer, chemist and mathematical physicist Josiah Williard Gibbs. He defined it as follows:

Gibbs noted also that for the purposes of this definition, any chemical element or combination of elements in given proportions may be considered a substance, whether capable or not of existing by itself as a homogeneous body. Chemical potential is also referred to as partial molar Gibbs energy (.

Colligative properties

Colligative properties are properties of solutions that depend on the number of molecules in a given volume of solvent and not on the properties (e.g. size or mass) of the molecules. Colligative properties include: lowering of vapor pressure; elevation of boiling point; depression of freezing point and osmotic pressure. Measurements of these properties for a dilute aqueous solution of a non-ionized solute such as urea or glucose can lead to accurate determinations of relative molecular masses.

Freezing-point depression	Freezing-point depression describes the phenomenon that the freezing point of a liquid (a solvent) is depressed when another compound is added, meaning that a solution has a lower freezing point than a pure solvent. This happens whenever a solute is added to a pure solvent, such as water. The phenomenon may be observed in sea water, which due to its salt content remains liquid at temperatures below $0\,°C$, the freezing point of pure water.
Phase	In the physical sciences, a Phase is a region of space (a thermodynamic system), throughout which all physical properties of a material are essentially uniform. Examples of physical properties include density, index of refraction, and chemical composition. A simple description is that a Phase is a region of material that is chemically uniform, physically distinct, and (often) mechanically separable.
Phase diagram	A Phase diagram in physical chemistry, engineering, mineralogy, and materials science is a type of chart used to show conditions at which thermodynamically-distinct phases can occur at equilibrium. In mathematics and physics, "Phase diagram" is used with a different meaning: a synonym for a phase space. Common components of a Phase diagram are lines of equilibrium or phase boundaries, which refer to lines that mark conditions under which multiple phases can coexist at equilibrium.
Freezing	In physical science, Freezing or solidification is the process in which a liquid turns into a solid when cold enough. The Freezing point is the temperature at which this happens. Melting, the process of turning a solid to a liquid, is almost the exact opposite of Freezing.
Ideal solution	In chemistry, an Ideal solution or ideal mixture is a solution in which the enthalpy of solution (or "enthalpy of mixing") is zero; the closer to zero the enthalpy of solution is, the more "ideal" the behavior of the solution becomes. Equivalently, an ideal mixture is one in which the activity coefficients (which measure deviation from ideality) are equal to one. The concept of an Ideal solution is fundamental to chemical thermodynamics and its applications, such as the use of colligative properties.
Cooling curve	A Cooling curve is a line graph that represents the change of phase of matter, typically from a gas to a solid or a liquid to a solid. The independent variable (X-axis) is time and the dependent variable (Y-axis) is temperature. Below is an example of a Cooling curve used in castings.
Dissociation	Dissociation in chemistry and biochemistry is a general process in which ionic compounds (complexes, molecules ions usually in a reversible manner. When a Bronsted-Lowry acid is put in water, a covalent bond between an electronegative atom and a hydrogen atom is broken by heterolytic fission, which gives a proton and a negative ion. Dissociation is the opposite of association and recombination.
Electrolyte	An Electrolyte is any substance containing free ions that behaves as an electrically conductive medium. Because they generally consist of ions in solution, Electrolyte s are also known as ionic solutions, but molten Electrolyte s and solid Electrolyte s are also possible.

Electrolyte s commonly exist as solutions of acids, bases or salts.

Equilibrium constant	Stability constants, formation constants, binding constants, association constants and dissociation constants are all types of Equilibrium constant.
Ionization	Ionization is the physical process of converting an atom or molecule into an ion by adding or removing charged particles such as electrons or other ions. This is often confused with dissociation (chemistry.)
	The process works slightly differently depending on whether an ion with a positive or a negative electric charge is being produced.
Activity	In chemical thermodynamics Activity is a measure of the "effective concentration" of a species in a mixture. By convention, it is a dimensionless quantity. The Activity of pure substances in condensed phases (solid or liquids) is normally taken as unity.
Ionic strength	The Ionic strength of a solution is a measure of the concentration of ions in that solution. Ionic compounds, when dissolved in water, dissociate into ions. The total electrolyte concentration in solution will affect important properties such as the dissociation or the solubility of different salts.
Liquid	Liquid is one of the principal states of matter. A Liquid is a fluid that has the particles loose and can freely form a distinct surface at the boundaries of its bulk material. The surface is a free surface where the Liquid is not constrained by a container.
Phase transition	In thermodynamics, a Phase transition is the transformation of a thermodynamic system from one phase to another.
	At a Phase transition point, physical properties may undergo abrupt change: for instance, the volume of the two phases may be vastly different as is illustrated by the boiling of liquid water to form steam.
	The term is most commonly used to describe transitions between solid, liquid and gaseous states of matter, in rare cases including plasma.
Vapor	A Vapor or vapour is a substance in the gas phase at a temperature lower than its critical temperature. This means that the Vapor can be condensed to a liquid or to a solid by increasing its pressure, without reducing the temperature.
	For example, water has a critical temperature of 374 °C (or 647 K) which is the highest temperature at which liquid water can exist.

Vapor pressure	Vapor pressure, is the pressure of a vapor in equilibrium with its non-vapor phases. All liquids and solids have a tendency to evaporate to a gaseous form, and all gases have a tendency to condense back into their original form At any given temperature, for a particular substance, there is a pressure at which the gas of that substance is in dynamic equilibrium with its liquid or solid forms.
Argon	Argon is a chemical element designated by the symbol Ar. Argon has atomic number 18 and is the third element in group 18 of the periodic table . Argon is present in the Earth"s atmosphere at 0.94%.
Vaporization	Vaporization of an element or compound is a phase transition from the liquid phase to gas phase. There are two types of Vaporization: evaporation and boiling. This diagram shows the nomenclature for the different phase transitions.
	Evaporation is a phase transition from the liquid phase to gas phase that occurs at temperatures below the boiling temperature at a given pressure.
Entropy	Entropy is a concept applied across physics, information theory, mathematics and other branches of science and engineering. The following definition is shared across all these fields:

$$S = -k \sum_i P_i \ln P_i$$

	where S is the conventional symbol for Entropy. The sum runs over all microstates consistent with the given macrostate and P_i is the probability of the ith microstate.
Distillation	Distillation is a method of separating mixtures based on differences in their volatilities in a boiling liquid mixture. Distillation is a unit operation, or a physical separation process, and not a chemical reaction.
	Commercially, Distillation has a number of uses.
Azeotrope	An Azeotrope is a mixture of two or more liquids in such a ratio that its composition cannot be changed by simple distillation. This occurs because, when an Azeotrope is boiled, the resulting vapor has the same ratio of constituents as the original mixture.
	Because their composition is unchanged by distillation, Azeotrope s are also called (especially in older texts) constant boiling mixtures.
Liquid crystals	Liquid crystals are substances that exhibit a phase of matter that has properties between those of a conventional liquid and those of a solid crystal. For instance, an Liquid crystals may flow like a liquid, but its molecules may be oriented in a crystal-like way. There are many different types of Liquid crystals phases, which can be distinguished by their different optical properties (such as birefringence.)

Order	Order in a crystal lattice is the arrangement of some property with respect to atomic positions. It arises in charge ordering, spin ordering, magnetic ordering, and compositional ordering. It is a thermodynamic entropy concept often displayed by a second Order phase transition.
Electrochemical cell	An Electrochemical cell is a device used for generating an electromotive force (voltage) and current from chemical reactions inducing a chemical reaction by a flow of current. The current is caused by the reactions releasing and accepting electrons at the different ends of a conductor. A common example of an Electrochemical cell is a standard 1.5-volt battery.
Critical point	In physical chemistry, thermodynamics, chemistry and condensed matter physics, a Critical point specifies the conditions at which a phase boundary ceases to exist. There are multiple types of critical points such as vapor-liquid critical points and liquid-liquid critical points. A plot of typical polymer solution phase behavior including two critical points: an LCST and a UCST. The liquid-liquid Critical point of a solution denotes the limit of the two-phase region of the phase diagram.
Critical opalescence	Critical opalescence is a phenomenon which arises in the region of a continuous phase transition. Originally reported by Thomas Andrews in 1869 for the liquid-gas transition in carbon dioxide, many other examples have been discovered since. The phenomenon is most commonly demonstrated in binary fluid mixtures, such as methanol and cyclohexane.

Electrolyte	An Electrolyte is any substance containing free ions that behaves as an electrically conductive medium. Because they generally consist of ions in solution, Electrolyte s are also known as ionic solutions, but molten Electrolyte s and solid Electrolyte s are also possible. Electrolyte s commonly exist as solutions of acids, bases or salts.
Solution	In chemistry, a Solution is a homogeneous mixture composed of two or more substances. In such a mixture, a solute is dissolved in another substance, known as a solvent. Gases may dissolve in liquids, for example, carbon dioxide or oxygen in water.
Chemical equilibrium	In a chemical process, Chemical equilibrium is the state in which the chemical activities or concentrations of the reactants and products have no net change over time. Usually, this would be the state that results when the forward chemical process proceeds at the same rate as their reverse reaction. The reaction rates of the forward and reverse reactions are generally not zero but, being equal, there are no net changes in any of the reactant or product concentrations.
Conductivity	The Conductivity of an electrolyte solution is a measure of its ability to conduct electricity. The SI unit of Conductivity is Siemens per metre (S/m.) Conductivity measurements are used routinely in many industrial and environmental applications as a fast, inexpensive and reliable way of measuring the ionic content in a solution.
Dissociation	Dissociation in chemistry and biochemistry is a general process in which ionic compounds (complexes, molecules ions usually in a reversible manner. When a Bronsted-Lowry acid is put in water, a covalent bond between an electronegative atom and a hydrogen atom is broken by heterolytic fission, which gives a proton and a negative ion. Dissociation is the opposite of association and recombination.
Strong electrolyte	A Strong electrolyte is a solute that completely ionizes or dissociates in a solution. These ions are good conductors of electric current in the solution. Originally, a "Strong electrolyte" was defined as a chemical that, when in aqueous solution, is a good conductor of electricity.
Equilibrium constant	Stability constants, formation constants, binding constants, association constants and dissociation constants are all types of Equilibrium constant.
Acid	An Acid is traditionally considered any chemical compound that, when dissolved in water, gives a solution with a hydrogen ion activity greater than in pure water, i.e. a pH less than 7.0. That approximates the modern definition of Johannes Nicolaus Brønsted and Martin Lowry, who independently defined an Acid as a compound which donates a hydrogen ion (H^+) to another compound (called a base.) Common examples include acetic Acid and sulfuric Acid (used in car batteries.)

Ionization	Ionization is the physical process of converting an atom or molecule into an ion by adding or removing charged particles such as electrons or other ions. This is often confused with dissociation (chemistry.) The process works slightly differently depending on whether an ion with a positive or a negative electric charge is being produced.
Phase	In the physical sciences, a Phase is a region of space (a thermodynamic system), throughout which all physical properties of a material are essentially uniform. Examples of physical properties include density, index of refraction, and chemical composition. A simple description is that a Phase is a region of material that is chemically uniform, physically distinct, and (often) mechanically separable.
Electrode	An Electrode is an electrical conductor used to make contact with a nonmetallic part of a circuit (e.g. a semiconductor, an electrolyte or a vacuum.) The word was coined by the scientist Michael Faraday from the Greek words elektron and hodos, a way. An Electrode in an electrochemical cell is referred to as either an anode or a cathode, words that were also coined by Faraday.
Hydrochloric acid	Hydrochloric acid is the solution of hydrogen chloride (HCl) in water. It is a highly corrosive, strong mineral acid and has major industrial uses. It is found naturally in gastric acid.
Enthalpy	In thermodynamics and molecular chemistry, the Enthalpy is a thermodynamic property of a fluid. It can be used to calculate the heat transfer during a quasistatic process taking place in a closed thermodynamic system under constant pressure. Enthalpy H is an arbitrary concept but the Enthalpy change ΔH is more useful because it is equal to the change in the internal energy of the system, plus the work that the system has done on its surroundings.
Heat	In physics and thermodynamics, Heat is the process of energy transfer from one body or system to another due to a difference in temperature. In thermodynamics, the quantity TdS is used as a representative measure of the (inexact) Heat differential δQ, which is the absolute temperature of an object multiplied by the differential quantity of a system"s entropy measured at the boundary of the object. A related term is thermal energy, loosely defined as the energy of a body that increases with its temperature.
Thermodynamic	In physics, Thermodynamic s ">power") is the study of the conversion of energy into work and heat and its relation to macroscopic variables such as temperature,volume and pressure. Its underpinnings, based upon statistical predictions of the collective motion of particles from their microscopic behavior, is the field of statistical Thermodynamic s, a branch of statistical mechanics. Historically, Thermodynamic s developed out of need to increase the efficiency of early steam engines.

Activity	In chemical thermodynamics Activity is a measure of the "effective concentration" of a species in a mixture. By convention, it is a dimensionless quantity. The Activity of pure substances in condensed phases (solid or liquids) is normally taken as unity.
Activity coefficient	An Activity coefficient is a factor used in thermodynamics to account for deviations from ideal behaviour in a mixture of chemical substances. In an ideal mixture the interactions between each pair of chemical species are the same (or more formally, the enthalpy of mixing is zero) and, as a result, properties of the mixtures can be expressed directly in terms of simple concentrations or partial pressures of the substances present e.g. Raoult"s law. Deviations from ideality are accommodated by modifying the concentration by an Activity coefficient.
Adsorption	Adsorption is the accumulation of atoms or molecules on the surface of a material. This process creates a film of the adsorbate (the molecules or atoms being accumulated) on the adsorbent"s surface. It is different from absorption, in which a substance diffuses into a liquid or solid to form a solution.
Electrochemical cell	An Electrochemical cell is a device used for generating an electromotive force (voltage) and current from chemical reactions inducing a chemical reaction by a flow of current. The current is caused by the reactions releasing and accepting electrons at the different ends of a conductor. A common example of an Electrochemical cell is a standard 1.5-volt battery.
Gas	In physics, a Gas is a state of matter, consisting of a collection of particles (molecules, atoms, ions, electrons, etc.) without a definite shape or volume that are in more or less random motion. Due to the electronic nature of the aforementioned particles, a "force field" is present throughout the space around them.
Ionic strength	The Ionic strength of a solution is a measure of the concentration of ions in that solution. Ionic compounds, when dissolved in water, dissociate into ions. The total electrolyte concentration in solution will affect important properties such as the dissociation or the solubility of different salts.
Platinum	Platinum is a chemical element with the chemical symbol Pt and an atomic number of 78." It is in Group 10 of the periodic table of elements. A dense, malleable, ductile, precious, gray-white transition metal, Platinum is resistant to corrosion and occurs in some nickel and copper ores along with some native deposits. Platinum is used in jewelry, laboratory equipment, electrical contacts and electrodes, Platinum resistance thermometers, dentistry equipment, and catalytic converters.
Order	Order in a crystal lattice is the arrangement of some property with respect to atomic positions. It arises in charge ordering, spin ordering, magnetic ordering, and compositional ordering. It is a thermodynamic entropy concept often displayed by a second Order phase transition.
Order of reaction	The Order of reaction , in chemical kinetics, with respect to a certain reactant, is defined as the power to which its concentration term in the rate equation is raised .

For example, given a chemical reaction 2A + B → C with a rate equation

$$r = k[A]^2[B]^1$$

the reaction order with respect to A would be 2 and with respect to B would be 1, the total reaction order would be 2+1=3. It is not necessary that the order of a reaction be a whole number - zero and fractional values of order are possible - but they tend to be integers.

Reaction mechanism	In chemistry, a Reaction mechanism is the step by step sequence of elementary reactions by which overall chemical change occurs .
	Although only the net chemical change is directly observable for most chemical reactions, experiments can often be designed that suggest the possible sequence of steps in a Reaction mechanism.
	A mechanism describes in detail exactly what takes place at each stage of a chemical transformation.
Activation	Activation in (bio-)chemical sciences generally refers to the process whereby something is prepared or excited for a subsequent reaction.
	In chemistry, Activation of molecules is required for a chemical reaction to occur. The phrase energy of Activation refers to the energy the reactants must acquire before they can successfully react with each other to produce the products, that is, to reach the transition state.
Activation energy	In chemistry, Activation energy is a term introduced in 1889 by the Swedish scientist Svante Arrhenius, that is defined as the energy that must be overcome in order for a chemical reaction to occur. Arrhenius" research was a follow up of the theories of reaction rate by Serbian physicist Nebojsa Lekovic. Activation energy may also be defined as the minimum energy required to start a chemical reaction.
Chemical kinetics	Chemical kinetics is the study of rates of chemical processes. Chemical kinetics includes investigations of how different experimental conditions can influence the speed of a chemical reaction and yield information about the reaction"s mechanism and transition states, as well as the construction of mathematical models that can describe the characteristics of a chemical reaction. In 1864, Peter Waage and Cato Guldberg pioneered the development of Chemical kinetics by formulating the law of mass action, which states that the speed of a chemical reaction is proportional to the quantity of the reacting substances.

Hydrolysis	Hydrolysis is a chemical reaction during which one or more water molecules are split into hydrogen and hydroxide ions, which may go on to participate in further reactions. It is the type of reaction that is used to break down certain polymers, especially those made by step-growth polymerization. Such polymer degradation is usually catalysed by either acid e.g. concentrated sulfuric acid (H_2SO_4) or alkali e.g. sodium hydroxide (NaOH) attack, often increasing with their strength or pH.

Hydrolysis is distinct from hydration, where the hydrated molecule does not "lyse" (break into two new compounds.) |
| Reversible reaction | A Reversible reaction is a chemical reaction that results in an equilibrium mixture of reactants and products. For a reaction involving two reactants and two products this can be expressed symbolically as

$$aA + bB \rightleftharpoons cC + dD$$

A and B can react to form C and D or, in the reverse reaction, C and D can react to form A and B. This is distinct from reversible process in thermodynamics.

The concentrations of reactants and products in an equilibrium mixture are determined by the analytical concentrations of the reagents (A and B or C and D) and the equilibrium constant, K. The magnitude of the equilibrium constant depends on the Gibbs free energy change for the reaction. |
| Catalysis | Catalysis is the process in which the rate of a chemical reaction is either increased or decreased by means of a chemical substance known as a catalyst. Unlike other reagents that participate in the chemical reaction, a catalyst is not consumed by the reaction itself. The catalyst may participate in multiple chemical transformations. |
| Relaxation | In nuclear magnetic resonance (NMR) spectroscopy and magnetic resonance imaging (MRI) the term Relaxation describes several processes by which nuclear magnetization prepared in a non-equilibrium state return to the equilibrium distribution. In other words, Relaxation describes how fast spins "forget" the direction in which they are oriented. The rates of this spin Relaxation can be measured in both spectroscopy and imaging applications. |
| Spin-spin relaxation time | Spin-spin relaxation time, known as T_2, is a time constant in nuclear magnetic resonance (NMR) and magnetic resonance imaging (MRI.) It is named in contrast to T_1, the spin-lattice relaxation time.

T_2 characterizes the rate at which the M_{xy} component of the magnetization vector decays in the transverse magnetic plane. |

Free induction decay	In Fourier Transform NMR, a Free induction decay is the observable NMR signal generated by non-equilibrium nuclear spin magnetisation precessing about the magnetic field (conventionally along z.) This non-equilibrium magnetisation is generally created by applying a pulse of resonant radio-frequency close to the Larmor frequency of the nuclear spins. If the magnetisation vector has a non-zero component in the xy plane, then the precessing magnetisation will induce a corresponding oscillating voltage in a detection coil surrounding the sample.
Chemical shift	In nuclear magnetic resonance (NMR), the Chemical shift describes the dependence of nuclear magnetic energy levels on the electronic environment in a molecule. Chemical shift s are relevant in NMR spectroscopy techniques such as proton NMR and carbon-13 NMR. An atomic nucleus can have a magnetic moment (nuclear spin), which gives rise to different energy levels and resonance frequencies in a magnetic field. The total magnetic field experienced by a nucleus includes local magnetic fields induced by currents of electrons in the molecular orbitals (note that electrons have a magnetic moment themselves.)
Enzymes	Enzymes are biomolecules that catalyze (i.e., increase the rates of) chemical reactions. Nearly all known Enzymes are proteins. However, certain RNA molecules can be effective biocatalysts too.
Enzyme kinetics	Enzyme kinetics is the study of the chemical reactions that are catalysed by enzymes, with a focus on their reaction rates. The study of an enzyme"s kinetics reveals the catalytic mechanism of this enzyme, its role in metabolism, how its activity is controlled, and how a drug or a poison might inhibit the enzyme. Enzymes are usually protein molecules that manipulate other molecules -- the enzymes" substrates.
Inversion	In meteorology, an Inversion is a deviation from the normal change of an atmospheric property with altitude. It almost always refers to a temperature Inversion, i.e., an increase in temperature with height, or to the layer (Inversion layer) within which such an increase occurs. An Inversion can lead to pollution such as smog being trapped close to the ground, with possible adverse effects on health.
Absorbance	In spectroscopy, the Absorbance A (also called optical density) is defined as $$A_\lambda = -\log_{10}(I/I_0),$$

where I is the intensity of light at a specified wavelength λ that has passed through a sample (transmitted light intensity) and I_0 is the intensity of the light before it enters the sample or incident light intensity. Absorbance measurements are often carried out in analytical chemistry, since the Absorbance of a sample is proportional to the thickness of the sample and the concentration of the absorbing species in the sample, in contrast to the transmittance I / I_0 of a sample, which varies logarithmically with thickness and concentration.

Outside the field of analytical chemistry, e.g. when used with the Tunable Diode Laser Absorption Spectroscopy (TDLAS) technique, the Absorbance is sometimes defined as the natural logarithm instead of the base-10 logarithm, i.e. as

$$A_\lambda = -\ln(I/I_0),$$

The term absorption refers to the physical process of absorbing light, while Absorbance refers to the mathematical quantity.

Activated complex	In chemistry an Activated complex is a transitional structure in a chemical reaction that results from an effective transfusion between molecules and that persists while unaccounted for bonds are breaking and new bonds are forming. It is therefore a range of molecular geometries along the reaction coordinate.
Pressure	Pressure is the force per unit area applied in a direction perpendicular to the surface of an object. Gauge Pressure is the Pressure relative to the local atmospheric or ambient Pressure.
	Pressure is an effect which occurs when a force is applied on a surface.
Cyclopentene	Cyclopentene is a chemical compound with the formula C_5H_8. It is a colorless liquid with a petrol-like odor. It is one of the cycloalkenes.
Half-life	The Half-life of a quantity whose value decreases with time is the interval required for the quantity to decay to half of its initial value. The concept originated in describing how long it takes atoms to undergo radioactive decay but also applies in a wide variety of other situations.
	The term "Half-life" dates to 1907.
Peroxide	A Peroxide is a compound containing an oxygen-oxygen single bond. The simplest stable Peroxide is hydrogen Peroxide Su Peroxide s, dioxygenyls, ozones and ozonides compound are considered separately.
Surface tension	Surface tension is an attractive property of the surface of a liquid. It is what causes the surface portion of liquid to be attracted to another surface, such as that of another portion of liquid (as in connecting bits of water or as in a drop of mercury that forms a cohesive ball.)

Applying Newtonian physics to the forces that arise due to Surface tension accurately predicts many liquid behaviors that are so commonplace that most people take them for granted.

| The Gibbs Adsorption Isotherm for Multicomponent systems | The Gibbs Adsorption Isotherm for Multicomponent systems |

The Gibbs Adsorption Isotherm is an equation used to relate the changes in concentration of a component in contact with a surface with changes in the surface tension. For a binary system containing two components the Gibbs Adsorption Equation in terms of surface excess is:

$$-\mathrm{d}\gamma = \Gamma_1 \mathrm{d}\mu_1 + \Gamma_2 \mathrm{d}\mu_2$$

where

γ is the surface tension
Γ_1 is the surface excess of component 1
μ_1 is the chemical potential of component 1

Different influences at the interface may cause changes in the composition of the near-surface layer Substances may either accumulate near the surface or conversely, move into the bulk. The movement of the molecules characterizes the phenomena of adsorption.

| BET theory | BET theory is a rule for the physical adsorption of gas molecules on a solid surface and serves as the basis for an important analysis technique for the measurement of the specific surface area of a material |

The concept of the theory is an extension of the Langmuir theory, which is a theory for monolayer molecular adsorption, to multilayer adsorption with the following hypotheses: (a) gas molecules physically adsorb on a solid in layers infinitely; (b) there is no interaction between each adsorption layer; and (c) the Langmuir theory can be applied to each layer.

| Liquid | Liquid is one of the principal states of matter. A Liquid is a fluid that has the particles loose and can freely form a distinct surface at the boundaries of its bulk material. The surface is a free surface where the Liquid is not constrained by a container. |

| Contact angle | The Contact angle is the angle at which a liquid/vapor interface meets the solid surface. The Contact angle is specific for any given system and is determined by the interactions across the three interfaces. Most often the concept is illustrated with a small liquid droplet resting on a flat horizontal solid surface. |

| Density | The Density of a material is defined as its mass per unit volume. The symbol of Density is ρ ">rho.) |

Mathematically:

$$\rho = \frac{m}{V}$$

where:

  ρ is the Density,
  m is the mass,
  V is the volume.

Density of air	The Density of air, ρ , is the mass per unit volume of Earth"s atmosphere, and is a useful value in aeronautics and other sciences. Air density decreases with increasing altitude, as does air pressure. It also changes with variances in temperature or humidity.

Compressibility

In thermodynamics and fluid mechanics, Compressibility is a measure of the relative volume change of a fluid or solid as a response to a pressure (or mean stress) change.

$$\beta = -\frac{1}{V}\frac{\partial V}{\partial p}$$

where V is volume and p is pressure. The above statement is incomplete, because for any object or system the magnitude of the Compressibility depends strongly on whether the process is adiabatic or isothermal.

Compressibility factor

The Compressibility factor is a useful thermodynamic property for modifying the ideal gas law to account for the real gas behaviour. In general, deviations from ideal behavior become more significant the closer a gas is to a phase change, the lower the temperature or the larger the pressure. Compressibility factor values are usually obtained by calculation from equations of state (EOS), such as the virial equation which take compound specific empirical constants as input.

Polymer

A Polymer is a large molecule composed of repeating structural units typically connected by covalent chemical bonds. While Polymer in popular usage suggests plastic, the term actually refers to a large class of natural and synthetic materials with a variety of properties.

Due to the extraordinary range of properties accessible in Polymer ic materials , they have come to play an essential and ubiquitous role in everyday life - from plastics and elastomers on the one hand to natural bio Polymer s such as DNA and proteins that are essential for life on the other.

Steric factor

Steric factor, P is a term used in collision theory.

It is defined as the ratio between the experimental value of the rate constant and the one predicted by collision theory. It can also be defined as the ratio between the preexponential factor and the collision frequency, and it is most often less than unity.

Gel

A Gel is a solid, jelly-like material that can have properties ranging from soft and weak to hard and tough. Gel s are defined as a substantially dilute crosslinked system, which exhibits no flow when in the steady-state. By weight, Gel s are mostly liquid, yet they behave like solids due to a three-dimensional crosslinked network within the liquid.

Viscometer

A Viscometer is an instrument used to measure the viscosity of a fluid. For liquids with viscosities which vary with flow conditions, an instrument called a rheometer is used. Viscometer s only measure under one flow condition.

Argon

Argon is a chemical element designated by the symbol Ar. Argon has atomic number 18 and is the third element in group 18 of the periodic table . Argon is present in the Earth"s atmosphere at 0.94%.

Solid

The Solid state of matter is characterized by a distinct structural rigidity and virtual resistance to deformation (i.e. changes of shape and/or volume.) Most Solid s have high values both of Young"s modulus and of the shear modulus of elasticity. This contrasts with most liquids or fluids, which have a low shear modulus, and typically exhibit the capacity for macroscopic viscous flow.

Energy levels

A quantum mechanical system or particle that is bound, confined spatially, can only take on certain discrete values of energy, as opposed to classical particles, which can have any energy. These values are called Energy levels. The term is most commonly used for the Energy levels of electrons in atoms or molecules, which are bound by the electric field of the nucleus.

Brownian motion

Brownian motion is the seemingly random movement of particles suspended in a fluid (i.e. a liquid or gas) or the mathematical model used to describe such random movements, often called a particle theory.

The mathematical model of Brownian motion has several real-world applications. An often quoted example is stock market fluctuations.

Dynamic light scattering

Dynamic light scattering is a technique in physics, which can be used to determine the size distribution profile of small particles in solution.

When light hits small particles the light scatters in all directions so long as the particles are small compared to the wavelength If the light source is a laser, and thus is monochromatic and coherent, then one observes a time-dependent fluctuation in the scattering intensity.

Polystyrene

Polystyrene), sometimes abbreviated PS, is an aromatic polymer made from the aromatic monomer styrene, a liquid hydrocarbon that is commercially manufactured from petroleum by the chemical industry. Polystyrene is one of the most widely used kinds of plastic.

Polystyrene is a thermoplastic substance, which is in solid (glassy) state at room temperature, but flows if heated above its glass transition temperature (for molding or extrusion), and becoming solid again when cooling off.

Phase transition

In thermodynamics, a Phase transition is the transformation of a thermodynamic system from one phase to another.

At a Phase transition point, physical properties may undergo abrupt change: for instance, the volume of the two phases may be vastly different as is illustrated by the boiling of liquid water to form steam.

The term is most commonly used to describe transitions between solid, liquid and gaseous states of matter, in rare cases including plasma.

Absorption

Absorption, in chemistry, is a physical or chemical phenomenon or a process in which atoms, molecules liquid or solid material. This is a different process from adsorption, since the molecules are taken up by the volume, not by surface. A more general term is sorption which covers adsorption, Absorption, and ion exchange.

Depolarization ratio

In Raman spectroscopy, the Depolarization ratio is the intensity ratio between the perpendicular component and the parallel component of the Raman scattered light.

The Raman scattered light is emitted by the stimulation of the electric field of the incident light. Therefore, the direction of the vibration of the electric field, or polarization direction, of the scattered light might be expected to be the same as that of the incident light.

Normal mode

A Normal mode of an oscillating system is a pattern of motion in which all parts of the system move sinusoidally with the same frequency. The frequencies of the Normal mode s of a system are known as its natural frequencies or resonant frequencies. A physical object, such as a building, bridge or molecule, has a set of Normal mode s that depend on its structure and composition.

Acetylene

Acetylene is the chemical compound with the formula HC_2H. It is a hydrocarbon and the simplest alkyne. This colourless gas is widely used as a fuel and a chemical building block. It is unstable in pure form and thus is usually handled as a solution.

Emission

In physics, Emission is the process by which the energy of a photon is released by another entity, for example, by an atom whose electrons make a transition between two electronic energy levels. The emitted energy is in the form of a photon.

The emittance of an object quantifies how much light is emitted by it.

Fluorescence	Fluorescence is a luminescence that is mostly found as an optical phenomenon in cold bodies, in which the molecular absorption of a photon triggers the emission of a photon with a longer (less energetic) wavelength, though a shorter wavelength emission is sometimes observed from multiple photon absorption. The energy difference between the absorbed and emitted photons ends up as molecular rotations, vibrations or heat. Sometimes the absorbed photon is in the ultraviolet range, and the emitted light is in the visible range, but this depends on the absorbance curve and Stokes shift of the particular fluorophore.
Resonance fluorescence	Resonance fluorescence is fluorescence from an atom or molecule in which the light emitted is at the same frequency as the light absorbed. In Resonance fluorescence, a photon is absorbed, causing an electron to jump to a higher level from which, after a delay, it falls back to its original level, emitting a photon having the same energy as the one absorbed. The emission direction is random.
Phosphorescence	Phosphorescence is a specific type of photoluminescence related to fluorescence. Unlike fluorescence, a phosphorescent material does not immediately re-emit the radiation it absorbs. The slower time scales of the re-emission are associated with "forbidden" energy state transitions in quantum mechanics.
Quenching	Quenching refers to any process which decreases the fluorescence intensity of a given substance. A variety of processes can result in Quenching, such as excited state reactions, energy transfer, complex-formation and collisional Quenching. As a consequence, Quenching is often heavily dependent on pressure and temperature.
Morse potential	The Morse potential is a convenient model for the potential energy of a diatomic molecule. It is a better approximation for the vibrational structure of the molecule than the quantum harmonic oscillator because it explicitly includes the effects of bond breaking, such as the existence of unbound states. It also accounts for the anharmonicity of real bonds and the non-zero transition probability for overtone and combination bands.
Sublimation	Sublimation of an element or compound is a transition from the solid to gas phase with no intermediate liquid stage. Sublimation is an endothermic phase transition that occurs at temperatures and pressures below the triple point At normal pressures, most chemical compounds and elements possess three different states at different temperatures.
Molecular orbital	In chemistry, a Molecular orbital is a mathematical function that describes the wave-like behavior of an electron in a molecule. This function can be used to calculate chemical and physical properties such as the probability of finding an electron in any specific region. The use of the term "orbital" was first used in English by Robert S. Mulliken in 1925 as the English translation of Schrödinger"s use of the German word, "Eigenfunktion".

Molecular orbital theory	In chemistry, Molecular orbital theory is a method for determining molecular structure in which electrons are not assigned to individual bonds between atoms, but are treated as moving under the influence of the nuclei in the whole molecule. In this theory, each molecule has a set of molecular orbitals, in which it is assumed that the molecular orbital wave function ψ_f may be written as a simple weighted sum of the n constituent atomic orbitals χ_i, according to the following equation:

$$\psi_j = \sum_{i=1}^{n} c_{ij} \chi_i$$

	The c_{ij} coefficients may be determined numerically by substitution of this equation into the Schrödinger equation and application of the variational principle. This method is called the linear combination of atomic orbitals approximation and is used in computational chemistry.
Spin-lattice relaxation time	Spin-lattice relaxation time, known as T_1, is a time constant in nuclear magnetic resonance and magnetic resonance imaging. It is named in contrast to T_2, the spin-spin relaxation time.
	T_1 characterizes the rate at which the longitudinal M_z component of the magnetization vector recovers.
Spectroscopy	Spectroscopy was originally the study of the interaction between radiation and matter as a function of wavelength (λ.) In fact, historically, Spectroscopy referred to the use of visible light dispersed according to its wavelength, e.g. by a prism. Later the concept was expanded greatly to comprise any measurement of a quantity as function of either wavelength or frequency.
Diffraction	Diffraction is normally taken to refer to various phenomena which occur when a wave encounters an obstacle. It is described as the apparent bending of waves around small obstacles and the spreading out of waves past small openings. Very similar effects are observed when there is an alteration in the properties of the medium in which the wave is travelling, for example a variation in refractive index for light waves or in acoustic impedance for sound waves and these can also be referred to as Diffraction effects.
Vapor	A Vapor or vapour is a substance in the gas phase at a temperature lower than its critical temperature. This means that the Vapor can be condensed to a liquid or to a solid by increasing its pressure, without reducing the temperature.
	For example, water has a critical temperature of 374 °C (or 647 K) which is the highest temperature at which liquid water can exist.

Vapor pressure	Vapor pressure, is the pressure of a vapor in equilibrium with its non-vapor phases. All liquids and solids have a tendency to evaporate to a gaseous form, and all gases have a tendency to condense back into their original form At any given temperature, for a particular substance, there is a pressure at which the gas of that substance is in dynamic equilibrium with its liquid or solid forms.
Antoine equation	The Antoine equation is a vapor pressure equation and describes the relation of the saturated vapor pressure and the temperature for pure components. The Antoine equation is derived from the Clausius-Clapeyron relation. $$P = 10^{A - \frac{B}{C+T}}$$ where P is the Pressure, T is the Temperature and A, B, C are component specific constants.
Chemical potential	Chemical potential, symbolized by μ, is a quantity first described by the American engineer, chemist and mathematical physicist Josiah Williard Gibbs. He defined it as follows: Gibbs noted also that for the purposes of this definition, any chemical element or combination of elements in given proportions may be considered a substance, whether capable or not of existing by itself as a homogeneous body. Chemical potential is also referred to as partial molar Gibbs energy (.
Entropy	Entropy is a concept applied across physics, information theory, mathematics and other branches of science and engineering. The following definition is shared across all these fields: $$S = -k \sum_i P_i \ln P_i$$ where S is the conventional symbol for Entropy. The sum runs over all microstates consistent with the given macrostate and P_i is the probability of the ith microstate.

Weston cell	The Weston cell, invented by Edward Weston in 1893, is a wet-chemical cell that produces a highly stable voltage suitable as a laboratory standard for calibration of voltmeters. It was adopted as the International Standard for EMF in 1911.
	The anode is an amalgam of cadmium with mercury, the cathode is of pure mercury, the electrolyte is a solution of cadmium sulfate octahydrate and the depolarizer is a paste of mercurous sulfate.
Activity	In chemical thermodynamics Activity is a measure of the "effective concentration" of a species in a mixture. By convention, it is a dimensionless quantity. The Activity of pure substances in condensed phases (solid or liquids) is normally taken as unity.
Activity coefficient	An Activity coefficient is a factor used in thermodynamics to account for deviations from ideal behaviour in a mixture of chemical substances. In an ideal mixture the interactions between each pair of chemical species are the same (or more formally, the enthalpy of mixing is zero) and, as a result, properties of the mixtures can be expressed directly in terms of simple concentrations or partial pressures of the substances present e.g. Raoult"s law. Deviations from ideality are accommodated by modifying the concentration by an Activity coefficient.
Gas	In physics, a Gas is a state of matter, consisting of a collection of particles (molecules, atoms, ions, electrons, etc.) without a definite shape or volume that are in more or less random motion.
	Due to the electronic nature of the aforementioned particles, a "force field" is present throughout the space around them.
Triple point	In thermodynamics, the Triple point of a substance is the temperature and pressure at which three phases (for example, gas, liquid, and solid) of that substance coexist in thermodynamic equilibrium. For example, the Triple point of mercury occurs at a temperature of –38.8344 °C and a pressure of 0.2 mPa.
	In addition to the Triple point between solid, liquid, and gas, there can be Triple point s involving more than one solid phase, for substances with multiple polymorphs.
Compressibility	In thermodynamics and fluid mechanics, Compressibility is a measure of the relative volume change of a fluid or solid as a response to a pressure (or mean stress) change.
	$$\beta = -\frac{1}{V}\frac{\partial V}{\partial p}$$
	where V is volume and p is pressure. The above statement is incomplete, because for any object or system the magnitude of the Compressibility depends strongly on whether the process is adiabatic or isothermal.

Compressibility factor	The Compressibility factor is a useful thermodynamic property for modifying the ideal gas law to account for the real gas behaviour. In general, deviations from ideal behavior become more significant the closer a gas is to a phase change, the lower the temperature or the larger the pressure. Compressibility factor values are usually obtained by calculation from equations of state (EOS), such as the virial equation which take compound specific empirical constants as input.
Platinum	Platinum is a chemical element with the chemical symbol Pt and an atomic number of 78." It is in Group 10 of the periodic table of elements. A dense, malleable, ductile, precious, gray-white transition metal, Platinum is resistant to corrosion and occurs in some nickel and copper ores along with some native deposits. Platinum is used in jewelry, laboratory equipment, electrical contacts and electrodes, Platinum resistance thermometers, dentistry equipment, and catalytic converters.
Density	The Density of a material is defined as its mass per unit volume. The symbol of Density is ρ ">rho.) Mathematically: $$\rho = \frac{m}{V}$$ where: ρ is the Density, m is the mass, V is the volume.
Liquid	Liquid is one of the principal states of matter. A Liquid is a fluid that has the particles loose and can freely form a distinct surface at the boundaries of its bulk material. The surface is a free surface where the Liquid is not constrained by a container.
Emission	In physics, Emission is the process by which the energy of a photon is released by another entity, for example, by an atom whose electrons make a transition between two electronic energy levels. The emitted energy is in the form of a photon. The emittance of an object quantifies how much light is emitted by it.
Emission spectrum	The Emission spectrum of an element or compound is the relative intensity of electromagnetic radiation of each frequency emitted by atoms or molecules of that element or compound when they are excited. Each atom"s atomic Emission spectrum is unique and can be used to determine if that element is part of an unknown compound. Similarly, the emission spectra of molecules can be used for chemical analysis.

Vapor	A Vapor or vapour is a substance in the gas phase at a temperature lower than its critical temperature. This means that the Vapor can be condensed to a liquid or to a solid by increasing its pressure, without reducing the temperature. For example, water has a critical temperature of 374°C (or 647 K) which is the highest temperature at which liquid water can exist.
Pressure	Pressure is the force per unit area applied in a direction perpendicular to the surface of an object. Gauge Pressure is the Pressure relative to the local atmospheric or ambient Pressure. Pressure is an effect which occurs when a force is applied on a surface.
Cold	ColD is a PLP-dependent enzyme responsible for the removal of the C-3" hydroxyl group during the biosynthesis of GDP-colitose.
Ion	An Ion is an atom or molecule where the total number of electrons is not equal to the total number of protons, giving it a net positive or negative electrical charge. Since protons are positively charged and electrons are negatively charged, if there are more electrons than protons, the atom or molecule will be negatively charged. This is called an an Ion , from the Greek á¼€vÎ¬ , meaning "up".
Density of air	The Density of air, ρ , is the mass per unit volume of Earth"s atmosphere, and is a useful value in aeronautics and other sciences. Air density decreases with increasing altitude, as does air pressure. It also changes with variances in temperature or humidity.
Ionization	Ionization is the physical process of converting an atom or molecule into an ion by adding or removing charged particles such as electrons or other ions. This is often confused with dissociation (chemistry.) The process works slightly differently depending on whether an ion with a positive or a negative electric charge is being produced.
Coefficient of thermal expansion	When the temperature of a substance changes, the energy that is stored in the intermolecular bonds between atoms changes. When the stored energy increases, so does the length of the molecular bonds. As a result, solids typically expand in response to heating and contract on cooling; this dimensional response to temperature change is expressed by its Coefficient of thermal expansion
PH meter	A PH meter is an electronic instrument used to measure the pH (acidity or alkalinity) of a liquid (though special probes are sometimes used to measure the pH of semi-solid substances.) A typical PH meter consists of a special measuring probe (a glass electrode) connected to an electronic meter that measures and displays the pH reading.

The pH probe measures pH as the activity of hydrogen ions surrounding a thin-walled glass bulb at its tip.

Electrode	An Electrode is an electrical conductor used to make contact with a nonmetallic part of a circuit (e.g. a semiconductor, an electrolyte or a vacuum.) The word was coined by the scientist Michael Faraday from the Greek words elektron and hodos, a way.
	An Electrode in an electrochemical cell is referred to as either an anode or a cathode, words that were also coined by Faraday.
Glass electrode	A Glass electrode is a type of ion-selective electrode made of a doped glass membrane that is sensitive to a specific ion.
	Almost all commercial electrodes responds to single charged ions, like H^+, Na^+, Ag^+. The most common Glass electrode is the pH-electrode.
Argon	Argon is a chemical element designated by the symbol Ar. Argon has atomic number 18 and is the third element in group 18 of the periodic table . Argon is present in the Earth"s atmosphere at 0.94%.
Doublet	In quantum mechanics, a doublet is a quantum state of a system with a spin of 1/2, such that there are two allowed values of the spin component, $-1/2$ and $+1/2$. Quantum systems with two possible states are sometimes called two-level systems. Essentially all occurrences of doublets in nature arise from rotational symmetry; spin 1/2 is associated with the fundamental representation of the Lie group SU(2), the group that defines rotational symmetry in three-dimensional space.
Heat	In physics and thermodynamics, Heat is the process of energy transfer from one body or system to another due to a difference in temperature. In thermodynamics, the quantity TdS is used as a representative measure of the (inexact) Heat differential δQ, which is the absolute temperature of an object multiplied by the differential quantity of a system"s entropy measured at the boundary of the object.
	A related term is thermal energy, loosely defined as the energy of a body that increases with its temperature.
Frequency spectrum	A source of light can have many colors mixed together and in different amounts (intensities.) A rainbow sends the different frequencies in different directions, making them individually visible at different angles. A graph of the intensity plotted against the frequency (showing the amount of each color) is the Frequency spectrum of the light.
Component	In thermodynamics, a Component is a chemically distinct constituent of a system. Calculating the number of components in a system is necessary, for example, when applying Gibbs phase rule in determination of the number of degrees of freedom of a system.

The number of components is equal to the number of independent chemical constituents, minus the number of chemical reactions between them, minus the number of any constraints (like charge neutrality or balance of molar quantities.)

Krypton

Krypton is a chemical element with the symbol Kr and atomic number 36. It is a member of Group 18 and Period 4 elements. A colorless, odorless, tasteless noble gas, Krypton occurs in trace amounts in the atmosphere, is isolated by fractionally distilling liquified air, and is often used with other rare gases in fluorescent lamps.

Neon

Neon is the chemical element that has the symbol Ne and atomic number 10. Although a very common element in the universe, it is rare on Earth. A colorless, inert noble gas under standard conditions, Neon gives a distinct reddish-orange glow when used in discharge tubes and Neon lamps.

Absorbance

In spectroscopy, the Absorbance A (also called optical density) is defined as

$$A_\lambda = -\log_{10}(I/I_0),$$

where I is the intensity of light at a specified wavelength λ that has passed through a sample (transmitted light intensity) and I_0 is the intensity of the light before it enters the sample or incident light intensity. Absorbance measurements are often carried out in analytical chemistry, since the Absorbance of a sample is proportional to the thickness of the sample and the concentration of the absorbing species in the sample, in contrast to the transmittance I / I_0 of a sample, which varies logarithmically with thickness and concentration.

Outside the field of analytical chemistry, e.g. when used with the Tunable Diode Laser Absorption Spectroscopy (TDLAS) technique, the Absorbance is sometimes defined as the natural logarithm instead of the base-10 logarithm, i.e. as

$$A_\lambda = -\ln(I/I_0),$$

The term absorption refers to the physical process of absorbing light, while Absorbance refers to the mathematical quantity.

Absorption

Absorption, in chemistry, is a physical or chemical phenomenon or a process in which atoms, molecules liquid or solid material. This is a different process from adsorption, since the molecules are taken up by the volume, not by surface. A more general term is sorption which covers adsorption, Absorption, and ion exchange.

Monochromator

A Monochromator is an optical device that transmits a mechanically selectable narrow band of wavelengths of light or other radiation chosen from a wider range of wavelengths available at the input. The name is from the Greek roots mono-, single, and chroma, colour, and the Latin suffix -ator, denoting an agent.

A device that can produce monochromatic light has many uses in science and in optics because many optical characteristics of a material are dependent on color.

Acetylene

Acetylene is the chemical compound with the formula HC_2H. It is a hydrocarbon and the simplest alkyne. This colourless gas is widely used as a fuel and a chemical building block. It is unstable in pure form and thus is usually handled as a solution.

Lightning Source UK Ltd.
Milton Keynes UK
01 July 2010

156338UK00001B/87/P